EVERY DAY IS EARTH DAY

HARRIET DYER

SIMPLE WAYS TO REDUCE YOUR CARBON FOOTPRINT

Andrews McMeel
PUBLISHING

EVERY DAY IS EARTH DAY

Andrews McMeel Publishing
a division of Andrews McMeel Universal
1130 Walnut Street, Kansas City, Missouri 64106

www.andrewsmcmeel.com

First published as *How to Reduce Your Carbon Footprint*
in 2020 by Summersdale Publishers Ltd.
Part of Octopus Publishing Group Limited
46 West Street
Chichester, West Sussex
PO19 1RP
United Kingdom

21 22 23 24 25 SDB 10 9 8 7 6 5 4 3 2 1

ISBN: 978-1-5248-6296-1

Library of Congress Control Number: 2020944160

Editor: Kevin Kotur
Art Director: Diane Marsh
Production Manager: Tamara Haus
Production Editor: Margaret Daniels

CONTENTS

Introduction iv

Weighing It Up vi

PLANET SOS: The Facts 1

HOW TO REDUCE YOUR CARBON FOOTPRINT 28

CLIMATE CHANGE CRUSADERS 111

BECOME AN ACTIVIST 116

Conclusion 119

INTRODUCTION

What is a carbon footprint?

We know that the main factor causing climate change today is the release of greenhouse gases, such as carbon dioxide (CO_2), into the atmosphere by human beings. But what is a carbon footprint? The term has gained a lot of traction over the past few years as an analogy when measuring the amount of greenhouse gases that people emit through various activities. Some activities have a carbon footprint that is created by one person, but others have a footprint that includes a complex chain of individuals and organizations.

Take, for example, a basic cotton T-shirt that was manufactured in another country. If you purchase it, at first glance the only emissions you might think you are accountable for are from traveling to and from the store (if you ride in a vehicle that emits CO_2). But the indirect carbon emissions (i.e., the ones that don't originate directly from you) that are linked to the T-shirt make your overall footprint a lot bigger. From the energy and water used to grow the raw material and manufacture the garment, to shipping the item and storing it in a warehouse or store, your indirect carbon footprint is the equivalent of approximately twenty pounds (or nearly ten kilograms)—think five two-liter bottles of soda—of CO_2 being released into the atmosphere.

Unfortunately, we live in a world where it's impossible not to leave a carbon footprint—even drinking water leaves a trace! However, understanding more about your direct and indirect carbon emissions will help you to make more informed decisions on how you can live a greener life.

What can I do?

With a problem that's so huge, it's easy to feel powerless and disheartened. But the truth is that small changes and actions can have big effects, because they all add up.

From finding out more about the crisis we are facing to learning ways in which you can make simple changes to your daily routine that'll make a big environmental difference, these pages will be your guide to a more eco-friendly life. Surprise your colleagues and acquaintances with carbon footprint facts, share the book with friends and family, and do everything in your power to get the message across that we must act now if we are going to help save future generations from climate catastrophe. There is still hope—so let's not give up the battle until we see some positive results . . . and then let's keep going.

WEIGHING IT UP

This book is quite data-heavy, but data can lose its impact if we don't have a way of picturing it. So, before we delve into the following chapters, here are some ways to visualize the big numbers you'll come across.

1 lb (.45 kg) = approx. two cups (roughly half a liter) of water

100 lb (45 kg) = approx. an adult chimpanzee

1 tn—or 2,000 lb (907 kg) = approx. one small car

150 tn (136,078 kg) = approx. a blue whale

1 acre (.41 hectares) = approx. ¾ of an American football field

1 gigaton—or 1,000,000,000 tn (two trillion pounds) = approx. six million blue whales

PLANET SOS: THE FACTS

" We are the first generation to feel the sting of climate change, and we are the last generation that can do something about it. **"**

Jay Inslee

In this chapter you can find out about the effects that our collective carbon footprint is having on the planet, from unusual climate and weather patterns, as well as changes to the atmosphere, to the rise of sea levels and forest depletion. It's your one-stop shop to discovering more about climate change.

OUR AIR AND ATMOSPHERE

We know it's there, but because it can't be seen with the naked eye, we often forget all about our atmosphere and the air that surrounds us and keeps us alive. Unfortunately, in this case, what's out of sight is out of mind. It's time to make the invisible visible as we look at some of the key facts and stats on how our massively oversized carbon footprint is changing the planet's climate, weather, and atmosphere.

A GLOBAL SNAPSHOT OF THE WEATHER AND CLIMATE

2019 was the second hottest year on record.

Hottest years on record (in order): 2016, 2019, 2015, 2017, 2018.

Since 1981, the global annual temperature has increased by 0.32°F (0.18°C) year.

At the time of writing, it is forecast to increase another two degrees within the next twenty to thirty years and over four degrees by 2100, on average.

Global precipitation has increased at an average rate of nearly an inch (2 cm) per decade since 1901.

The number of floods and periods of heavy rain has quadrupled since 1980 and doubled since 2004.

In North America, the length of time when snow covers the ground has become shorter by nearly two weeks since 1972. In the Northern Hemisphere, snow cover has decreased by 5 percent annually, from 1966 to 2005.

USA

In 2019, there were over 120,000 extreme weather records, including snow in Hawaii and heatwaves in Alaska.

The country had rainfall 4.7 in (12 cm) above average in 2019.

The year 2018 was ranked fourth for most weather disasters in a year, behind 2017 (sixteen events), 2011 (sixteen) and 2016 (fifteen). It had two tropical cyclones, eight severe storms, two winter storms, drought, and wildfires.

Asia

The year 2019 was Asia's third hottest on record.

There were 213 heatwaves from 1980 to 1999 in India. Between 2000 and 2018, in roughly the same period of time, there were 1,400.

The year 2019 saw a mean annual temperature in Singapore of 83.1°F (28.4°C) and is the country's joint warmest year on record (also 2016).

This century has witnessed nine of Singapore's ten hottest years.

Temperatures in Iran hit 129.2°F (54°C) in 2017 (one of the hottest temperatures ever recorded).

Europe

The number of warm days have doubled between 1960 and 2018.

Number of extreme heatwaves in Europe since 2000: nine (2003, 2006, 2007, 2010, 2014, 2015, 2017, 2018, and 2019).

France, Czech Republic, Slovakia, Austria, Andorra, Luxembourg, Poland, and Germany experienced record temperatures in June 2019.

Australia

Jan–Feb 2019: A total of 206 weather records were broken over ninety days, including: record-high summer temperature in eighty-seven locations, record-low summer total rainfall in ninety-six locations, and record-high summer rainfall in fifteen places.

The heat caused mass deaths of wild horses, native bats, and fish in drought-affected areas.

Bushfires in Tasmania burned 500,000 acres (over 200,000 hectares) of vegetation.

Townsville, Queensland, received 4.1 ft (1.257 m) of rain over ten days in late January and early February 2019—more than its annual average rainfall.

Jun 2019–Mar 2020: Bushfires swept across more than 46 million acres (over 18.7 million hectares) of Australia and killed over a billion animals.

Approximately 337 million tons (over 674 billion pounds) of CO_2 were emitted during this emergency.

After the height of the bushfires, Sydney experienced 15.4 in. (391.6 mm) of rain, the most rainfall since 1990.

Africa

In 2019, Mozambique was hit by a record two cyclones in a single season.

Somalia is suffering from long-term drought:

- Famine killed roughly 258,000 people between 2010 and 2012.
- It failed to rain for three seasons in a row in 2016 and 2017.
- Over a period of forty-eight hours in 2017, 110 people died from starvation and drought-related illness.
- During the 2018 rainy season, the region received just 25–50 percent of average rainfall.
- Over 137,000 people were forced to flee their homes in the first quarter of 2019.

Long-term droughts (from 2015–18) in Cape Town, South Africa, resulted in water levels in dams being 15–30 percent below their total capacity. Huge water restrictions were put in place to prevent an emergency.

OUR ATMOSPHERE

Our atmosphere is essential to sustaining life on Earth. It's one of the most important things we should be protecting, yet you could say we're being pretty bad caretakers. Let's have a look at the damage we are causing.

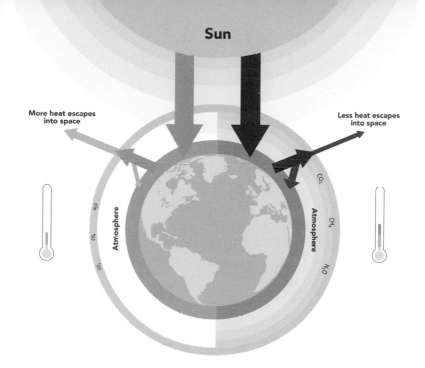

CLIMATE CHANGE EXPLAINED

The greenhouse gases that are released into the atmosphere create a blanket between the Sun and Earth, so that the Sun's heat gets trapped in the Earth's atmosphere. The left-hand side shows a planet with small amounts of greenhouse gasses in its atmosphere, and the right-hand side shows what happens when we have a larger amount.

In the 1950s, the levels of atmospheric CO_2 were 300–310 ppm (that's parts per million, out of all the molecules in the air). In 2019, CO_2 concentration of 415.26 ppm was detected. The last time Earth's atmosphere contained this much CO_2 was more than three million years ago.

Human activity is responsible for contributing as much as 26.4 gigatons—over fourteen trillion pounds (think back to the earlier chart and that equates to the mass of 160 million blue whales)—of CO_2 *per year* into the atmosphere.

Around half of the CO_2 emitted since 1750 is from the past forty years.

Over a period of 20–200 years, 65–80 percent of CO_2 that's been released into the atmosphere is dissolved into the ocean. The remaining molecules take thousands of years to disappear.

OUR OCEANS

Our oceans cover 71 percent of the Earth's surface. They play a huge part in life on our planet, but they are at risk due to a staggering amount of plastic ending up in them, the death of reefs because of global warming, the increase in acidification from the chemicals we use in farming, and the loss of many marine species due to commercial fishing. Let's have a look at how they have changed over the years.

TEN LOWEST MAXIMUM ARCTIC SEA ICE EXTENTS 1979–2019

	Year	In millions of mi² (km²)
1	2017	5.56 (14.41)
2	2018	5.59 (14.48)
3	2016	5.60 (14.51)
4	2015	5.61 (14.52)
5	2011	5.66 (14.67)
6	2006	5.67 (14.68)
7	2007	5.70 (14.77)
8	2019	5.71 (14.78)
9	2005	5.77 (14.95)
10	2014	5.78 (14.96)

Since 1880, over the last 140 years, global sea levels have risen by about 9 in. (23 cm). In the past twenty-five years alone, they have increased approx. 3.1 in. (8 cm).

Sea levels are currently rising at a rate of 0.1 in. (3.6 mm) per year. This is two and a half times quicker than the average rate of rising sea levels during most of the twentieth century.

Sea levels could rise another 43.3 in. (1.1 m) by 2100. If this happens, around 2 billion people—or about a fifth of the world's population—will be forced to move further inland.

OUR REEFS

The frequency of coral-reef bleaching (the loss of life-giving algae in the coral tissue, as a result of warming oceans) has increased by five times compared to 40 years ago.

Scientists have claimed that more than 90 percent of corals will die by 2050 (even if we miraculously managed to stop climate change).

More than 75 percent of the Earth's tropical reefs experienced bleaching-level heat stress between 2014 and 2017. In 2016 and 2017, this affected 900 mi (1,450 km) of coral along the 1,430 mi (2,300 km) Great Barrier Reef.

What are the consequences of coral reefs disappearing?

- There will be no spawning or feeding ground for important marine life.

- A unique habitat that supports an entire ecosystem will be lost.

- Certain medicines will be impossible to produce without key ingredients found in the reefs.

OUR LAND

The Earth's landscape has changed pretty dramatically over the past two hundred years. While whole forests are being removed and glaciers are melting, landfills are expanding and humans are multiplying, adding to the amount of CO_2 released into the atmosphere. Let's get a glimpse of how we are changing the land on our planet.

OUR FORESTS

Approximately thirty million acres (about 120,000 km²) of carbon-absorbing tropical forest was cut down in 2018—the equivalent of thirty football fields per minute.

In the past two decades, Afghanistan has lost over 70 percent of its forests.

Across the world, half of all rainforests have been destroyed. It is estimated that, at the current rate, they will disappear completely within the next one hundred years.

Deforestation is now responsible for 10 percent of the world's greenhouse gas emissions: not only does it minimize a natural carbon filter, it actually releases CO_2 when trees are cleared or burned.

Up to 28,000 species are expected to become extinct by 2050 due to deforestation.

OUR WASTELAND

Landfills produce lots of gas—90–98 percent of which is carbon dioxide and methane—as bacteria tries to decompose waste in an unnatural environment.

In 2016, cities around the world generated 2.2 billion tons (over two trillion kg) of waste.

With the expected increase in the human population, we will see annual worldwide waste grow to 3.7 billion tons (over 3.4 trillion kg) by 2050.

In developing countries, over 90 percent of waste is burned in the open air or dumped illegally.

By 2022, England's current landfill sites will exceed maximum capacity.

At the time of writing, America has 3,091 active landfills, generates 30 percent of the world's waste, and yet makes up just 5 percent of the world's population.

Four biggest dumping grounds in the world (2014):

Estrutural, Brasilia, Brazil: size of 194 football fields

Duquesa, Santo Domingo, Dominican Republic: size of 183 football fields

Bantar Gebang, Jakarta, Indonesia: size of 160 football fields

Truitier, Port-au-Prince, Haiti: size of 134 football fields

Total: 671 football fields' worth of waste

TOP SEVEN
DISAPPEARING GLACIERS

The Alps: Nearly half the glaciers have disappeared since record-keeping began.

Muir Glacier, Alaska: Between 1941 and 2004, the glacier retreated more than 7.5 mi (12 km) and thinned by 2,600 ft (over 800 m).

Himalayas: The annual loss of ice from the year 2000 to the present has doubled compared to the years 1975–2000.

Helheim Glacier, Greenland: In 2000–2005 the glacier retreated 4.5 mi (7.2 km).

Mount Kilimanjaro: The snows have melted more than 80 percent since 1912. Nearby, the Lewis Glacier, Mount Kenya, has lost nearly ALL of its ice at an average of 40 in. (1 m) per year.

In Greenland, the daily total of ice melt in July 2019 was 153,499 mi^2 (397,560 km^2), an 18% growth from the average 1988–2017 melt (which is 130,481 mi^2—or 337,945 km^2—per day).

Affected areas in Antarctica are losing ice five times faster than in the 1990s, with more than 328 ft (100 m) of thickness gone in some places.

The Andes: In the 1990s, scientists predicted that the Chacaltaya glacier would disappear by 2015, but this claim was dismissed as an overreaction. The glacier actually disappeared in 2009.

Montana's Glacier National Park:
The number of glaciers has declined to fewer than 30, from more than 150 in the late nineteenth century. It is predicted that almost all the ice in this region will disappear by 2030.

Antarctica has contributed to approximately 0.3 in. (7 mm) of sea level rise since 1992.

PLANET OF THE HUMANS

Our technological and medical advances have allowed us to live longer and have more children than ever. The human population keeps rising, which, in turn, means that more and more carbon is being released.

HUMAN POPULATION OVER THE PAST 2,000 YEARS

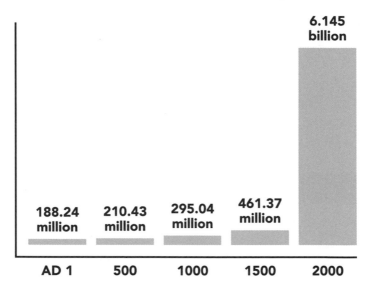

Did you know?

It has been reported that having one fewer child per family will save an average of 64.6 tons (58,600 kg) of CO_2-equivalent emissions per year.

THE ANIMAL KINGDOM

As we clear forests for agriculture, wild animals are often killed or forced to leave their natural habitats and so become vulnerable. Not only are we reducing the size of our planet's carbon filter and releasing CO_2 into the atmosphere from felling trees, but we are also greatly harming the animal kingdom.

60 percent

- The current percentage of wild animal populations that have been wiped out since 1970.

- The current percentage of mammals that are livestock.

At the time of writing, just 4 percent of the mammal population is wild; 36 percent is humans.

70 percent

- The current percentage of birds that are farmed poultry.

In the past five hundred years, human activity has forced over eight hundred species into extinction.

At the time of writing, over 26,000 of the world's species are endangered.

A SNAPSHOT OF WILD-ANIMAL POPULATIONS VS. DOMESTICATED-ANIMAL POPULATIONS

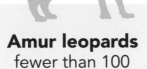

Amur leopards
fewer than 100

Horses
58 million

Tigers
3,900

Cats
600 million

Masai giraffes
35,000

Dogs
900 million

Blue whales
10,000–25,000

Pigs
678 million

Orangutans
fewer than 120,000

Chickens
23 billion

HOW TO REDUCE YOUR CARBON FOOTPRINT

> **"**We cannot solve a crisis without treating it as a crisis.**"**
>
> **Greta Thunberg**

- -

If anything major is to be done about climate change, we need to pressure those most responsible for it: global corporations that see targets and profits as the be-all and end-all. But let's not forget that if every individual on the planet trod a little more lightly, then we might be able to start saving our beautiful home and send a message to those in power.

Before we explore how we can reduce our carbon footprint in all areas of our lives, let's have a look at which global sectors are the biggest climate change offenders:

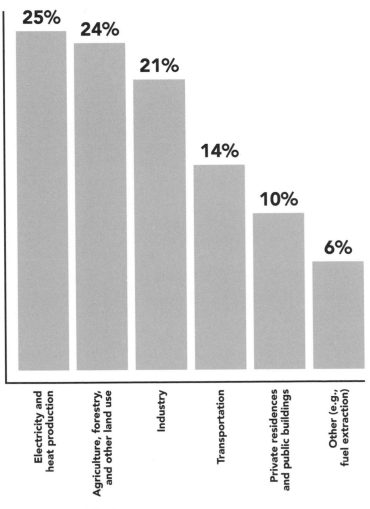

	25%	24%	21%	14%	10%	6%
	Electricity and heat production	Agriculture, forestry, and other land use	Industry	Transportation	Private residences and public buildings	Other (e.g., fuel extraction)

ENERGY AT HOME

The average home in the West uses up energy constantly, even when not occupied. Here's a chart to show the appliances that consume the most energy:

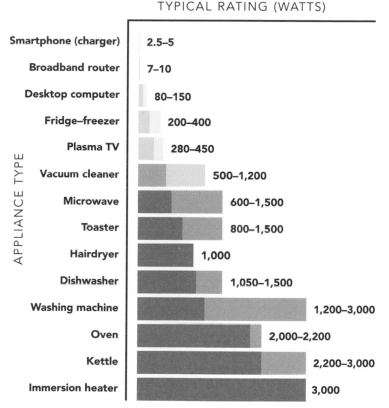

TYPICAL RATING (WATTS)

Appliance type	Typical rating (watts)
Smartphone (charger)	2.5–5
Broadband router	7–10
Desktop computer	80–150
Fridge–freezer	200–400
Plasma TV	280–450
Vacuum cleaner	500–1,200
Microwave	600–1,500
Toaster	800–1,500
Hairdryer	1,000
Dishwasher	1,050–1,500
Washing machine	1,200–3,000
Oven	2,000–2,200
Kettle	2,200–3,000
Immersion heater	3,000

USE LESS ELECTRICITY AND GAS BY SWITCHING TO:

Low power and screen brightness settings on your electronic devices, such as laptops and phones, so you don't have to charge them as much.

A washing line outside or clotheshorse in a well-ventilated room, rather than using a tumble dryer.

Did you know?

One cycle of a tumble dryer is equivalent to turning on 225 light bulbs for an hour.

If every UK household with a tumble dryer dried one load of washing naturally per week, collectively they would save over a million tons (nearly one billion kilograms) of CO_2 every year.

Using a microwave or toaster oven for pre-made or small meals, instead of a gas or electric oven—this can reduce energy consumption by 80 percent.

Washing dishes by hand.

- Use a washing tub and only fill with enough water to fully submerge plates.
- Install a low-flow aerator to reduce the amount of rinsing water you use.
- Leave the washing of dirty kitchenware till last so that you can make a fresh bowl of water go further.

In some areas, you can receive a green-energy tax credit by switching to:

A smart meter, which will help you track exactly how much energy you use at home. Some energy suppliers will install this free of charge.

Did you know?

In the UK, 62 percent of homeowners who have a smart meter say that they are now using their energy at home more wisely.

Compact fluorescent lamps (CFLs) or light-emitting diodes (LEDs), which use approx. 25–80 percent less energy and last up to 25 times longer than traditional light bulbs.

Having radiator reflectors behind your radiators to help the circulation of heat in rooms.

Appliances that are more energy efficient (but only when your current ones are on their last legs or you want to sell them secondhand).

An energy-efficient convection gas cooker, which will cut cooking times by up to 30 percent.

- Don't pre-heat your oven for longer than necessary.
- Open the oven door as little as possible.
- Use the right-sized pans for burners to eliminate heat waste.
- Put lids on pans to boil water and cook food quicker.

Using a laptop or tablet rather than a desktop computer.

Thick, lined curtains to reduce heat escaping through your windows.

Heating timers and a thermostat to control when the heat should be on.

All of the following switches have big initial upfront costs, but investing in them will help you really reduce your carbon footprint and save money on bills in the long run.

Underfloor heating

Solar panels

Heat pumps

Cavity wall insulation

Loft insulation (to cut costs, install the insulation yourself)

Double- or triple-glazed windows

BECOME A MORE MINDFUL ECO-WARRIOR AND START:

Turning lights off when a room isn't in use.

Charging your gadgets in the day (instead of overnight), unplugging them as soon as they are fully charged, and turning plug switches off.

Did you know?

Every year in the United States, $19 billion of energy is wasted from fully charged devices being plugged in.

Standby mode accounts for up to 10 percent of residential power use and the biggest energy drainers are TV sets.

Did you know?

People in the UK waste £68 million (approx. $85 million) a year in energy by boiling more water than they need in their kettles.

Using the correct amount of water you need when boiling water in a kettle.

Sealing window and door frames—you can reapply the seal yourself when you notice it becoming thinner.

Using draught excluders to prevent cold air getting inside.

Washing your clothes with cold water and using a quicker time setting.

Using rechargeable batteries or properly recycling old single-use batteries.

Did you know?

When batteries go to landfills, they still have dangerous leftover chemicals inside them.

When the cases of the batteries break down, the chemicals are released into the soil and enter the environment.

Moving furniture away from air vents and radiators so heat is properly circulated.

Turning radiators off or switching vents closed in rooms that aren't used.

Shutting all doors in the colder months to keep in the heat.

Wearing extra layers of clothes or snuggling up in blankets, if you're cold.

WATER WASTE

Perhaps you're wondering how using water adds to the size of your carbon footprint? It's not all that obvious at first, so here's a summary of the process of industrial-scale water treatment to ensure it is clean when we use it:

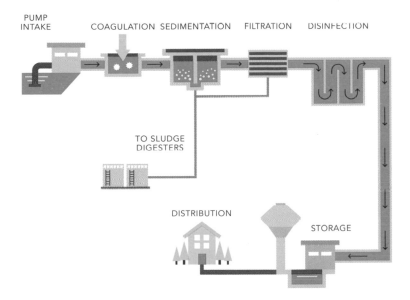

PUMP INTAKE   COAGULATION SEDIMENTATION   FILTRATION   DISINFECTION

TO SLUDGE DIGESTERS

DISTRIBUTION

STORAGE

All this machinery needs energy in order to make it run and we get most of our energy by burning fossil fuels. The water industry contributes to 3–7 percent of total greenhouse gas emissions.

HOUSEHOLD WATER USE

It's easy to fall into the trap of using more water than is required and, since we use it here and there, it's very difficult to visualize how much of it we are wasting in the long term. Let's look at the household appliances that typically use the most water:

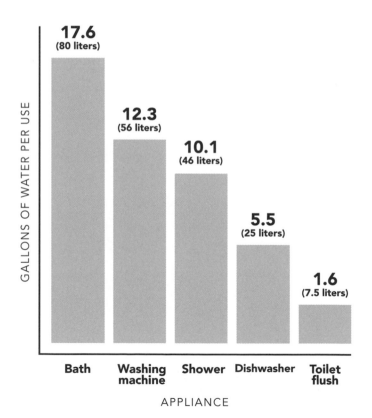

Although, at first glance, it seems that flushing the toilet isn't so bad, think about how many times you flush a day. And then take that figure and multiply it by 3 billion (approximately the amount of people in the world who have access to flushable toilets). It actually ends up being the biggest contributor of household water use. Perhaps it's time to put the saying "If it's yellow, let it mellow" into practice?

Did you know?

In the US, the average use of water is 100 gallons (455 liters) per person per day. That adds up to 36,500 gallons (138,000 liters) per person per year—over three times what the average citizen in the UK uses.

Most of the household water we use is heated—around 30 percent of our energy bills come from this—and further contributes to the burning of fossil fuels.

BECOME A MORE CONSCIOUS ECO-WARRIOR AND REDUCE YOUR WATER USE BY:

Taking showers instead of baths.

Replacing worn washers on leaky faucets.

Did you know?

Every minute spent in the shower uses approx. 1.8 gallons (8 liters) of water. Cutting the time you spend in the shower by one minute each day means that you'll stop using up to 12.3 gallons (56 liters) of water per week.

Did you know?

Up to 16.9 gallons (77 liters) of household water per day is wasted due to leaks in pipes, faucets, and appliances.

Using your washing machine less—it's OK to wear clothes more than once without washing them if they don't stink!

Using eco-programs on your appliances.

Wash clothes only when you have a full load.

Installing a toilet with a dual flush button and using it correctly.

Washing dishes all at once, with one bowl of water, rather than periodically throughout the day.

Turning off the faucet when brushing your teeth.

Connecting a rain barrel to your roof drainage and using the contents for cleaning your car, watering plants, washing your windows, etc.

Filling your sink to the required amount for cleaning your razor, washing fruits and vegetables, cleaning dishes, etc. instead of running them under a faucet.

Did you know?

By 2025, over 2.8 billion people across 48 countries will face problems accessing enough water.

CLEANING

Buying cleaning products in stores is very convenient, but this convenience unfortunately comes at a cost to the planet. Here are just some of the reasons they smell fresh, but stink at the same time:

- The chemicals in them end up in our water cycle.

- They are typically packaged in plastic.

- The factories that make them consume lots of energy.

- The supply chain—getting the products from the factory to your home—contributes to burning fossil fuels.

So the next time you do your weekly shopping, instead of stocking up on cleaning products, make your own just-as-effective, cheaper, and planet-friendly soaps and sprays.

**To make
washing powder:**
Use equal amounts of
grated bars of soap
and baking soda,
plus a few drops of
essential oil to make it
smell fresh.

**To make
surface cleaner:**
Use equal amounts of
water and white vinegar,
adding the juice of
half a lemon and a few
drops of eucalyptus or
tea tree oil.

To make dish soap:
Use equal amounts of white
vinegar and water, and add the
juice of half a lemon.

To make heavy-duty cleaner:
Mix baking soda and
water to make a paste.

THE CARBON FOOTPRINT OF PLASTIC

Plastic is a huge part of the carbon cycle; in fact, 8 percent of the oil production worldwide is used to manufacture plastic. In terms of its usefulness, the lifespan of plastic can be very short (think of single-use plastics like drinks bottles) and what is sent to landfills each year is enough to circle the world four times. As shown in the table to the right, unlike other man-made materials, plastic doesn't biodegrade, so it becomes a pollutant if it escapes into nature.

Did you know?

Plastic is not only littering the world, it is also killing animals and marine life, as well as increasing the risk of cancer in humans. Studies that have looked into the dangers of plastic suggest that some of the chemicals, such as bisphenol A (BPA), at certain exposure levels, may be absorbed into the water or earth around it (i.e., the chemicals from a plastic bottle can be absorbed into the drinking water) and cause cancer.

MATERIAL	TIME TAKEN TO BIODEGRADE
Paper	2–5 months
Cotton T-shirt	6 months
Wool socks	1–5 years
Leather shoes	25–40 years
Nylon fabric	30–40 years
Tin cans	50–100 years
Aluminum cans	80–100 years
Styrofoam cup	500 years to forever
Plastic bags	500 years to forever
Glass bottles	1 million years

HOW TO REDUCE YOUR PLASTIC CARBON FOOTPRINT

Packaging is often the biggest plastic culprit, but alternatives are readily available, if you look in the right places.

Bathroom products and toiletries

Instead of liquids, choose solid products that require no/little packaging, such as soap, shampoo, conditioner and deodorant bars, toothpaste tabs, and bath bombs.

Avoid products that contain microbeads and microplastics—on most labels they are called either "polypropylene" or "polyethylene"—which are mainly found in health and beauty products, such as toothpaste and facial scrubs.

Use more paper-, cotton-, and bamboo-based products, such as cotton buds with paper stems, bamboo toothbrushes (if your gums and teeth aren't sensitive), and cotton and bamboo pads.

Swap to reusable sanitary products, or 100-percent cotton tampons or pads.

Opt for makeup that is packaged in wooden or tin containers.

Ditch the disposable razors.

Buy coconut oil in bulk and use it to remove make-up, or as a deodorant, moisturizer, and conditioner (see below for other uses).

Store away small, empty containers to fill with homemade toiletries into when you go on vacation.

Cleaning products

Opt for recycled toilet paper packaged in paper.

Use carbolic soap as a surface cleaner.

Use one to two cups of white vinegar to clean the toilet and use a toilet brush to scrub off any stubborn stains.

Use two tablespoons of coconut oil mixed with two tablespoons of baking soda to remove stains.

Reuse cleaning cloths, boiling them in water after each use, until they wear out.

Swap sponges for bamboo scrubbers.

Keep old clothes, towels, and bed sheets to use as cloths, and toothbrushes to use for cleaning off grime and mold.

Use washing powder and soda crystals that come in cardboard packaging.

Kitchen items

Use beeswax wraps instead of cling film for preserving food.

Buy loose fruits and vegetables.

Buy bakery items that come in paper bags.

Buy pantry items in bulk to maximize the food-to-packaging ratio.

Make a conscious effort to buy food and drink items that are packaged in easier-to-recycle glass or aluminum.

Buy loose-leaf tea or tea bags that are biodegradable—be careful, as most tea bags contain plastic.

Stay clear of items packaged in a significant amount of single-use plastic, such as coffee pods and microwave meals.

Buy long-lasting kitchenware, such as stainless steel for the cooktop and clay and stoneware for the oven.

RECYCLING

You want to recycle more, but it sometimes feels like the goalposts of what is required of you keep changing. Is it OK to put things in the recycling without cleaning them? What can be recycled and what can't? Is it true that if a soiled item goes in with the rest, then that whole batch goes to a landfill? And what about the process of transporting and actually recycling things—doesn't this just offset the good?

There are many recycling skeptics out there, so let's look at the facts and stats to see why recycling is our friend, not our foe:

It **saves 10–15 million tons (over 10 billion kilograms) of CO_2 a year**—the equivalent of taking 3.5 million cars off roads.

Recycling aluminum uses **95 percent less energy** than making it from raw materials.

Recycling paper uses **40 percent less energy** than making it from scratch, and for every 1 ton (907 kilograms) of paper that is recycled, seventeen trees are saved.

By recycling plastic, **70 percent less plastic** is created.

It **stops more trash from going to landfills,** which a) are almost at maximum capacity, and b) cost governments a lot of money—it is twice as expensive to send waste to landfills than to recycle it.

For every **one job in waste management and disposal, four jobs in the recycling industry are required,** thus helping to create more jobs.

To know that your recycling isn't going to be a wasted effort, be aware of the following:

- You **must** clean your products before putting them in your recycling bin, but that doesn't mean wasting water by rinsing them individually. Collect all your dirty recycling and wash it with the water left in the sink after you've washed the dishes.

- Recycling differs between regions, as some facilities are more equipped and advanced than others. Therefore, it's always best to check your local authority's website to see what they accept. Generally, plastic containers, paper, cardboard, glass, and aluminum can all be recycled.

- Shredded paper and plastic bottle lids are fine to recycle (but some local authorities don't accept them, so always check on their website first).

- Newspaper and cardboard boxes used to package takeout food generally absorb too much grease to be recycled. It is best to throw these in a compost bin (see p. 114 on how to start one) or general waste.

Did you know?

A global company called TerraCycle runs a waste management system for hard-to-recycle materials (such as dessert and chip packets, cigarette butts, and razors), which you drop off at specific locations and they recycle them into something else. Visit www.terracycle.com for more information.

FOOD AND DRINK

Since the advent of globalization, it's been very easy for developed countries to get hold of huge varieties of food, sometimes at the cost of exploiting poorer countries. Since our food travels far, it is shipped in layers and layers of shrink-wrap and plastic to prevent it being damaged.

Food-to-go and instant meals are the worst offenders in terms of packaging, as each serving is wrapped individually (and then again) and sometimes also collectively in a multipack. Meat, fish, and dairy products, on the other hand, win the jackpot for biggest contributors to climate change. Read on to find out why. . . .

Livestock accounts for 14.5 percent of all human-caused greenhouse gas emissions, almost two-thirds of which are from cattle.

83 percent of global farmland is dedicated to raising livestock.

An average healthy meat-heavy diet creates carbon dioxide emissions of 15.9 lb (7.2 kg) per day, while an average healthy vegan diet contributes just 6.4 lb (2.9 kg) per day.

Farming is accountable for 81 percent of the food supply chain's greenhouse gas emissions, 79 percent of ocean acidification (the decrease in the pH levels in the sea, which harms its wildlife), and 95 percent of eutrophication (run-off of minerals and nutrients into bodies of water, which causes a nutrient imbalance).

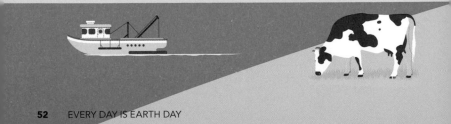

In 2016, 228 million tons (roughly 207 billion kilograms) of CO_2 was released into the atmosphere by marine fishing vessels. This is approximately equivalent to the amount of CO_2 emitted by 51 coal-fired power plants over the same period of time.

Commercial fishing is diminishing marine populations—the total number of Pacific bluefin tuna is 97 percent smaller than historic levels.

Now let's take a look at exactly how many greenhouse gas emissions are released—from production to getting it onto our plates, and everything in between—by the processes behind some of our most common foods:

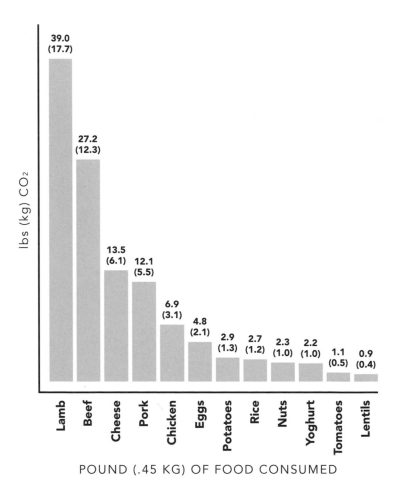

SAYING NO TO MEAT, FISH, AND DAIRY

Studies show that switching to a plant-based—not just vegetarian—diet is the single biggest way to reduce your impact on climate change. But if meat, eggs, and dairy have been a significant part of your diet, then it might seem difficult to flick the switch and cut them out of your life completely. Try taking small steps to a meat-, egg-, and dairy-free life by:

Adding vegan sides to the meals you'd usually have to begin with, in order to make the transition feel less overwhelming. This way, you'll be able to decide which bits you love, like, and dislike before making whole platefuls.

Switching from cow's milk to dairy-free milk (soy, oat, etc.), which is one of the easiest steps to veganism. Most people agree that soy goes well in tea, and oat and almond complement cereal and coffee.

Making a plan for how you want to transition to vegan food. Cut meat and dairy out of your diet one food item at a time and stick with it for at least a week before you eliminate the next thing. Start with the food you don't find all that appetizing, like that ham sandwich you have for lunch every day, and swap it for something vegan that you find much tastier, be it a superfood salad or lentil curry.

Creating vegan dishes with a familiar flavor to what you might be used to tasting. Most meat gets its flavor from the spices and seasoning used to cook or marinate it. Try, for example, adding pre-made fajita spice mix to black beans instead of chicken when you have your next Mexican feast.

Checking the back of packets, bags, and boxes of the food you enjoy to see if they are vegan. Going vegan doesn't necessarily mean a total overhaul of everything you love eating. You might be surprised to find out that most "chicken" and "beef" instant noodles are actually meat-free!

Starting with quick and easy vegan dishes once you've decided to ditch the meat. Getting stuck into complex and time-consuming dishes at first might put you off the prospect altogether. Once you've mastered some basic meals and want to try more complicated ones, be prepared to allow a bit more time than you would when cooking meat dishes.

Opting for "real" vegan dishes rather than meat and cheese substitutes. Although vegan meat and cheese look uncannily the same as the real stuff, you'll notice they don't have the same taste and texture—and they are often quite expensive.

Writing down your reasons for eating more vegan food— or cutting out meat and dairy products for good—and displaying that list somewhere in your kitchen to encourage you to stick at it.

Jotting down your favorite vegan recipes in a notepad. This will help you to decide what your routine meals are.

If cutting out all these foods is proving difficult, look at the chart on p. 60 to see which ones have the biggest impact on greenhouse gas emissions—lamb, beef, cheese, pork—and opt to remove them from your diet. You could even try the flexitarian diet, where you have meat or fish as a treat.

FEEDING THE PLANET

In order to feed the billions of us on the planet, the food and drink industry is growing at around 5 percent a year; it is estimated that consumer expenditure will reach $20 trillion by 2030.

> ## Did you know?
>
> Each year, about one-third of food (approx. $990 billion) gets lost or is wasted and generates more than 3.6 billion tons (over 3.3 trillion kilograms) of excess CO_2.

A significant amount of energy goes into every single step of getting food onto shelves: from growing and harvesting the crops, to making the products, to packaging them and delivering them to a store. On the following pages is a breakdown of these steps and how they impact our planet.

THERE'S SOMETHING IN THE WATER

This is something that probably doesn't even cross our minds when we think about how our food and drink (but also most things that are manufactured) are produced: "virtual water," which is the volume of water required to make the things we consume. From coffee and meat to milk and fruit, the chart below indicates how many gallons (or liters) of water per pound of product that we are using indirectly when we buy certain foodstuffs:

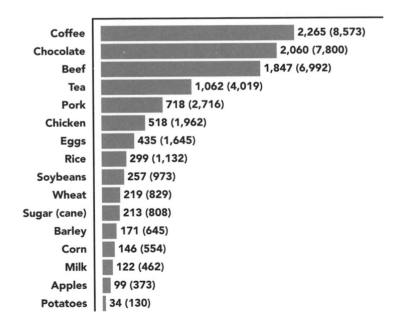

Coffee	2,265 (8,573)
Chocolate	2,060 (7,800)
Beef	1,847 (6,992)
Tea	1,062 (4,019)
Pork	718 (2,716)
Chicken	518 (1,962)
Eggs	435 (1,645)
Rice	299 (1,132)
Soybeans	257 (973)
Wheat	219 (829)
Sugar (cane)	213 (808)
Barley	171 (645)
Corn	146 (554)
Milk	122 (462)
Apples	99 (373)
Potatoes	34 (130)

A LONG WAY FROM HOME

Whereas people used to be restricted to buying food only grown in the country they lived in, now we can pick up any specialized ingredient from across the globe at our local supermarket. Not only is everything more readily available to us, but high profit margins set out by businesses also mean that crops and livestock are often sourced from developing countries, usually a very long way from the country of import. The table on the next page gives an idea of where the world's common foodstuffs are most likely to come from (by country, 2018).

PRODUCT	FIRST	SECOND	THIRD	FOURTH	FIFTH
Chicken	China	Indonesia	United States	Brazil	Iran
Cattle	Brazil	India	United States	China	Ethiopia
Bananas	India	China	Indonesia	Brazil	Ecuador
Oranges	Brazil	China	India	United States	Mexico
Apples	China	United States	Poland	Turkey	India
Strawberries	China	United States	Mexico	Egypt	Turkey
Cauliflower and broccoli	China	India	United States	Mexico	Spain
Potatoes	China	India	Ukraine	Russia	United States
Carrots and turnips	China	Uzbekistan	United States	Russia	Ukraine
Sugar cane	Brazil	India	China	Thailand	Pakistan
Maize (corn)	United States	China	Brazil	Argentina	Ukraine
Oats	Russia	Canada	Spain	Australia	Poland
Rice, paddy	China	India	Indonesia	Bangladesh	Vietnam
Cocoa	Ivory Coast	Ghana	Indonesia	Nigeria	Cameroon
Eggs	China	United States	India	Mexico	Brazil
Honey	China	Turkey	Argentina	Iran	Ukraine
Black pepper	Vietnam	Brazil	Indonesia	India	Bulgaria

HOW MANY MILES DOES YOUR FOOD HAVE TO TRAVEL BEFORE IT GETS TO YOUR PLATE?

OK, so let's say that you live in the UK and you enjoy oats and honey for breakfast, eggs for lunch, and a dinner of chicken, potatoes, and broccoli. Let's see how many miles this typical daily menu might have to travel (as the crow flies and using the table on the previous page) before you get to eat it.

Honey ➡ 4,831 mi (7,775 km)
Oats ➡ 3,486 mi (5,610 km)
= 8,317 mi (13,385 km)

Eggs
= 4,831 mi (7,775 km)

Chicken ➡ 4,831 mi (7,775 km)
Potatoes ➡ 4,831 mi (7,775 km)
Broccoli ➡ 4,831 mi (7,775 km)
= 14,493 mi (23,324 km)

**Grand total =
27,641 mi (44,484 km)**

WAYS TO REDUCE YOUR FOOD CARBON FOOTPRINT

Although getting the groceries done in one go at the local supermarket (where food comes from far and wide) is a widespread habit, there are things everyone can do to try to reduce their food carbon footprint:

Reduce the amount of "luxury" products you have. These are things, such as coffee, tea, chocolate, and sugar, that aren't the nutritional crux of your survival. You'll feel healthier for it, too!

Check the country of origin info on food packaging and stickers, and **always opt for the item that has traveled the shortest distance.** If choice is limited or non-existent, consider finding a supermarket that sources and sells local produce.

Base your meals on produce that's in season so that you can buy more locally sourced items.

Shop at independent stores rather than big supermarkets: You'll be helping to reduce emissions while also doing your part to save local businesses.

Sign up for a produce box subscription: These companies try to make sure that the fruits and vegetables are locally sourced and mainly seasonal.

Grow your own (see p. 113): From a vegetable patch in the garden to windowsill herbs and spices, growing your own is the perfect opportunity to really reduce those food miles.

Only buy fruit and vegetables that are loose to reduce the amount of plastic packaging that you throw away.

Look at the packaging of produce you usually buy and **choose the brand that uses the least materials** or more eco-friendly ones (i.e., not plastic).

Buy food that has a long shelf life in bulk and store it in large glass containers.

Don't get tempted by promotional offers. **Buy only what you need**—or even buy less than what you think you need and make it last!

Cut down on portion sizes and get rid of snacks. It's easier to do than you might think.

Implement a zero-waste policy at home and create forfeits for when the rules are broken, such as cooking dinner/washing the dishes for the week. You'll probably find you save quite a bit on your food bill by doing this.

Buy food that keeps: If you notice you're buying salad and only using half the bag before it starts going slimy, in the future avoid it and anything else that has a short shelf life. Instead, hardy produce, such as root vegetables, will last ages—and even when it starts sprouting, you can easily pick the bits off.

"Best before" isn't the same as "use by"—it just means that the quality of the food is best before the date shown, but it can still be eaten up until you see/smell the signs that it has gone bad.

Use your leftovers: Save cooked vegetables you don't eat for your next meal, make banana bread from overripe bananas, and blend fruit that's spoiled to create smoothies; use stale bread for breadcrumbs and croutons; add dinner leftovers to soup, stir-fries, casseroles, chili, wraps, omelettes, and frittatas; freeze or preserve, where possible.

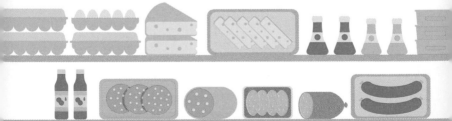

SHOPPING

It is estimated that, by 2030, five billion people worldwide will belong to the "consumer class," a type of lifestyle revolving around accumulating non-essential goods. Considering this, it's clear to see that we have a slight obsession with buying material things. If we're going to stop climate change in its tracks, we need to reevaluate our habits and focus on purchasing fewer but better items that are sustainably and ethically made.

THE CARBON FOOTPRINT OF CLOTHES

The fashion industry is hazardous to the Earth, accounting for 8 percent of global climate change. "Fast fashion," in particular, has a detrimental impact on the environment: it is cheap, mass-produced (usually unethically), poorly made with low-quality synthetic fabrics, and has a limited longevity. Yet we still buy these products, only to complain months down the line that they have shrunk or faded or have holes in them.

Let's take a look at some hidden truths surrounding the fast-fashion industry:

It produces 20 percent of global waste water.

It contributes to 10 percent of the world's carbon emissions.

Each second, the equivalent of one garbage truck of textiles is incinerated or taken to landfills.

Fabric dye is the second biggest polluter of water.

In 2017, the equivalent of over $15 million of clothes were thrown away and 300,000 tons (approx. 270 million kilograms) of textiles were sent to landfills.

WAYS TO BUY LESS FAST FASHION

Try alternative fabrics: Advances mean natural fiber types, such as bamboo, hemp, and nettle, now perform well compared to the more widely used textiles. Some brands are also recycling water bottles into polyester for clothing—just look on the labels for "100-percent recycled polyester" or "made with partially recycled materials."

Choose clothes that fit and suit you: Don't buy (and ultimately discard) clothes that you might wear only once, might soon go out of fashion, or "were an absolute bargain." Ask yourself: will I wear this repeatedly, do I need it, and does it complement what I have already?

Support smaller independent manufacturers and retailers.

Buy brands that accept their products back for repair, if damaged, and that are willing to recycle if garments are returned to them after years of use.

Avoid online shopping: For more info, see pages 95–98.

Favor the "make do and mend" mentality: Rework existing clothes and fabrics into new ones, or fix wear and tear. Broken zips or fasteners can be easily replaced; if you don't want to do it yourself, consider taking the garment to a tailor.

Look for vintage finds: Vintage clothes are considered better in quality and, if in their original state, are often made of natural fibers untouched by modern chemicals, dyes, and solvents.

Create a capsule wardrobe, i.e., a small collection of high-quality, simple, versatile, and long-lasting clothes, shoes, and accessories, all of which could be ethically and environmentally sourced. Take into consideration:

- The number of items you'll limit yourself to—somewhere around the region of 30–50 is practical.

- The color palette—ensure the items are interchangeable and complementary.

- What you can afford—you want clothes that are good quality, and that will come at a price you probably aren't used to. However, they are guaranteed to last longer, meaning that in the long term you are saving money. Do your research on companies that are renowned for sustainability before you start purchasing items.

- Don't dump all the clothes you already own to start a capsule wardrobe—instead, in order to be as sustainable as possible, you can phase out garments once they break beyond repair or are looking shabby.

THE CARBON FOOTPRINT OF ELECTRONICS

Electronics are a staple of modern life, but with gadgets ranging from smart speakers to electronic trash cans, all too often the novelty of the product outweighs the necessity. And this comes at a significant cost to the planet. . . .

Data centers worldwide, which keep the internet running, use 3–5 percent of total consumed electricity, rivaling the aviation industry's carbon footprint.

A typical smartphone requires 75 lb (34 kg) of rock to be mined for just 3.5 oz (100 g) of minerals.

The virtual currency Bitcoin produces enough carbon dioxide per year to complete 1 million transatlantic flights.

Using a smartphone for one hour a day for a year is the equivalent of a round-trip flight between New York City and Chicago (approx. 1,466 mi, or 2,360 km).

Each year, we produce enough e-waste (discarded electrical or electronic devices) to bury the equivalent of San Francisco (46.9 mi^2, or 121.5 km^2) in 13.7 ft (4.2 m) of it.

It is estimated that 70 percent of the human population owns a smartphone. These devices have an average user span (i.e., being owned by one person) of just two years.

Just one fifth of e-waste is recycled. The rest is burned, creating harmful gases and chemicals.

HOW TO CURB YOUR ELECTRONIC CARBON FOOTPRINT

Don't be lured into phone upgrades: By signing up to a monthly contract you have the opportunity to frequently upgrade. Save your money and the environment by holding onto your device until it breaks. If you buy the phone outright and choose a SIM-only deal, you'll save a lot of money in the long term and be less tempted to always have the latest model.

If you need a new phone, **buy refurbished rather than new**.

Switching off cell phones, laptops, and computers when you aren't using them may help to extend their lives.

Buy a laptop or computer model that can be easily upgraded and repaired.

Anything that has a plug or rechargeable battery needs to be taken to a recycling center to be properly recycled.

Upgrade the software on your computer or laptop rather than buying a new machine. This can save around 551 lb (250 kg) of fossil fuels.

Commit to not buying fad electronic devices. They are easy gifts to give when you don't know what to buy for a friend or family member but more often than not, novelty presents are relegated to the back of the closet.

Go electronic cold turkey: See how much you miss being connected to the virtual world by switching off all devices for a week, and only giving yourself half an hour a day to catch up on emails and texts. If you like it, make it a permanent routine.

Make small changes in your life to ditch the devices. Instead of chatting on apps, meet up in real life more frequently; get your entertainment fix by regaling people with events from your day rather than watching funny (most likely staged) videos online; choose a board or card game, read a book, go for a walk, or enjoy crafting instead of watching TV.

THE CARBON FOOTPRINT OF TOYS AND BABY PRODUCTS

We want our children to have the best, but we often forget how this impacts the planet. From shiny plastic toys to disposable diapers, the stats tell a surprising story.

In 2018, £3.3 billion (over $4 billion) was spent on 371 million toys in the UK.

In the US, a child receives on average sixty-nine new toys per year.

Studies carried out by a UK charity found that children have an average of four toys that they've never even played with.

On average, Americans spend $306 (about £240) on toys per child per year. Brits spend £159 (approx. $200) on toys in a baby's first year and £208 (approx. $255) in a baby's second year.

Each year in the UK, around 8.5 million like-new toys are thrown away.

Approximately 20 billion disposable diapers are sent to landfills in the US each year; in the UK, 3 billion are thrown away.

Throughout their first years, up to when they are potty trained, the average baby goes through 4,000–6,000 disposable diapers.

In Britain, 11 billion baby wipes are used per year.

WAYS TO BE BOTH CHILD- AND ECO-FRIENDLY

To be extra eco-friendly, opt for reusable diapers for a smaller carbon footprint: You'll be creating much less waste, but be mindful of how you clean them. Instead of using the baby-clothes setting on your washing machine, which is energy-intensive, try to select the lowest temperature option that still cleans them effectively. You can also soak them in hot, soapy water in a bucket first to lift the majority of the stains. When drying them, do this on the line in a sunny spot to bleach them white.

Replace single-use wet wipes that take hundreds of years to decompose with more eco-friendly alternatives. Reusable wipes are cheaper, better for baby and don't go to a landfill after one use. Majority opinion is to use bamboo, rather than fleece, for a better clean.

Buy or accept secondhand clothes and toys: Accept hand-me-downs or look on online marketplaces for secondhand (most often like-new/only-worn-once) bargains. There is no need to buy brand-new.

Think twice before buying: Too often toys and clothes are bought and then end up collecting dust. If you have a very young child, think about what they actually need. Request more eco-friendly, useful presents from family and friends for birthdays and Christmas—or ask that they don't buy anything at all.

Try toy libraries: Look up and register to a local toy library and enjoy the benefits of borrowing toys as you would books.

THE CARBON FOOTPRINT OF THE CHRISTMAS HOLIDAY

Although Santa is usually the one deciding who's been naughty and nice, it's about time we reassess the situation and find out how naughty Santa is being to our planet.

Over the two main days of Christmas it's reported that the average person uses 57 lb (26 kg) of CO_2 in food, 212 lb (96 kg) in travel, 481 lb (218 kg) in lighting, and 683 lb (310 kg) in consumables. That's 5.5 percent of the average person's annual carbon footprint.

During the holiday season every year, Americans spend roughly $6 billion on wrapping paper—more money than the entire economy of some countries—and go through 4 million pounds (1.8 million kilograms) of it, most of which ends up in the trash.

The average American also spends $942 dollars on holiday gifts, for a total of more than $1 trillion annually.

The carbon footprint of buying a real Christmas tree (if it doesn't end up in a landfill) is the equivalent of driving 12 mi (19 km) in a car; the carbon footprint of buying an artificial tree is driving 135 mi (217 km), although artificial trees do last longer.

In the UK in 2014, 4.2 million Christmas dinners were reportedly wasted. This amounts to 263,000 turkeys, 17.2 million Brussels sprouts, 11.9 million carrots, 11.3 million roast potatoes, and 9.8 million cups of gravy.

WAYS TO REDUCE YOUR CARBON FOOTPRINT AT CHRISTMAS

Decorating:

Rent a tree: There are companies that allow you to rent a Christmas tree for a period of time and then, when the festivities end, they take it back and replant it. Conveniently, a lot of these companies will deliver the tree to your door, too.

Replant your Christmas tree: If you have bought a young, potted Christmas tree and it's still healthy, you can replant it in your backyard (if you have one) and dig it up again the following year.

Buy LED lights instead of incandescent lights, as they consume a fraction of the energy.

Only turn on your lights when it gets dark and remember to turn them off before you go to sleep.

Craft your own homemade decorations: Make paper chains and snowflakes with the kids, go foraging and dehydrate fruit for potpourri materials, frost pine cones for tree ornaments, create your own natural wreath, and assemble a centerpiece using citrus fruits, cranberries, and spices.

Make a reusable advent calendar that you can refill each year with different surprises.

Gift giving:

Organize a Secret Santa rather than buying individual presents for everyone.

Avoid buying gift sets, as they are heavily packaged with single-use materials.

Opt for presents that are natural, e.g., paraffin-free candles, 100-percent organic cotton clothes or cushions, homemade soap, reusable water bottles, potted plants or flowers, homemade baked goods and condiments, bee's wrap sandwich wraps, repurposed chopping boards, and knitted garments.

Buy gifts that are "experiences," such as an annual nature reserve membership or a cooking course, but bear in mind what the experience is and where it is, as sometimes they can leave a worse footprint than physical presents.

Use brown paper or newspaper, and string or paper tape, for wrapping presents.

Buy cards that are FSC® certified and don't have special finishes, such as foil or glitter, or let people know that instead of sending Christmas cards you'll donate that money to a charity.

Food:

Use uneaten turkey, chicken, ham, or other meat as the base for your next dinners; you can make curry, casserole, pie, fajitas—whatever you'd like.

Turn your leftover vegetables, such as carrots and potatoes, into a soup.

Use your potatoes for potato cakes or a frittata.

Add leftover cranberry sauce to your granola and yogurt bowl, dollop it on top of pancakes, or use it to make cranberry muffins.

Go eggnog-crazy and use your leftovers to make eggnog cheesecake, cupcakes, rice pudding, brownies, truffles, and French toast.

Put your unused sauce or gravy in the fridge and reheat it for your next dinner (maybe for your casserole). Or you can fill ice-cube trays with the gravy and pop them in the freezer to make stock cubes.

Work out the amount of food that each person you are cooking for will eat— the shops are only closed for one day, if that, so buy less, plan smaller portions, and make it last.

THE CARBON FOOTPRINT OF FURNITURE

It has been suggested that for every piece of furniture that is made, an average of 104 lb (47 kg) of carbon dioxide and its equivalents are emitted into the atmosphere. Sofas are the worst culprits, with the average generating 198 lb (90 kg) of pollutants:

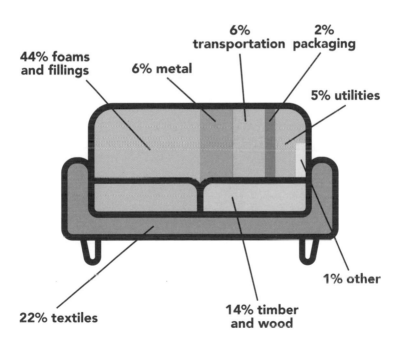

6% transportation   **2% packaging**

44% foams and fillings   **6% metal**

5% utilities

22% textiles

14% timber and wood

1% other

Here are other estimates of the average amount of CO_2 generated from making furniture:

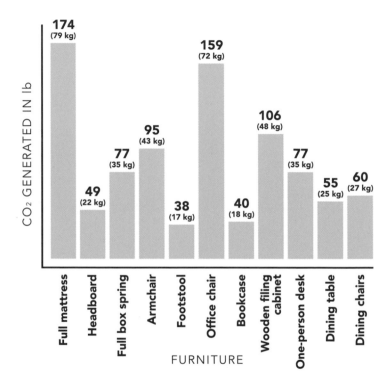

Did you know?

In 2018, Americans spent $111 billion dollars on furniture and bedding. Meanwhile, 9.8 million tons (8.9 billion kilograms) of furniture are sent to US landfills every year.

FURNISH YOUR HOME THE ECO-FRIENDLY WAY

It's lovely to make your house a home you can be proud of, but before you head straight to IKEA, try to be more mindful of the footprint you are leaving. Here are some tips on how to beautify your home sustainably:

Buy secondhand furniture: Charity/thrift stores and, of course, secondhand furniture stores—as well as online marketplaces and auction sites—are great for this. Older furniture can often be sturdier than modern stuff. Try to go for solid wood pieces and don't worry if there are a few marks or chips on them, since you can treat them quite easily. There are also a number of ways to upcycle wood furniture, from sanding to bare natural wood to varnishing and painting. There are many online tutorials on how to do this.

If you want to buy new, don't buy cheap: Don't waste your money on flimsy plywood that won't last very long. Instead, make your furniture a real investment and spring for (as much as you can afford to) good-quality, solid pieces that have a long guarantee on them. Also, check whether they are "green"— i.e., they are made with sustainable or reclaimed materials, have low toxic material levels, and are manufactured locally.

Always check for the FSC® logo to make sure the material has been sourced sustainably.

Purchase furniture and furnishings that you'll be happy with until they are on their last legs: Sometimes it's best not to go for statement pieces that you might start to dislike after a couple of years. Instead, opt for timeless pieces that you can accessorize. Here's how long most household items should last (provided they are good-quality brands):

> Mattress—at least ten years
> Carpet—at least 20 years
> Sofa—at least 15 years (leather sofas can last three times as long)
> Wood furniture—a lifetime

Buy antique furniture at an auction: Despite most people assuming that any description with the word "antique" in it will be way out of their price range, this furniture can be reasonable, especially if bought at an auction, and the quality is guaranteed.

Did you know?

A new chest of drawers that is manufactured in China has a carbon footprint sixteen times higher than its antique equivalent.

Take the correct steps to dispose of your unwanted furniture: Firstly, see if you can sell it locally and, if not, contact a secondhand store or furniture charity to see if they'll accept it. Alternatively, if you don't have a vehicle or one that's big enough to take the item to your local recycling center, contact your local authority to see if they do a bulk waste collection. This will cost a small fee but at least your conscience will be free of guilt.

THE TRAFFIC-LIGHT CHOICES OF ONLINE SHOPPING

In the past, we would have had to traipse downtown for what we needed, but now everything and anything can be bought at the click of a button. Online shopping has encouraged us to favor cheap prices over quality, to want impossibly fast delivery and free returns, and to make impulsive—often unnecessary—purchases because of clever marketing and the enormous selection on offer. But how do all these contribute to climate change?

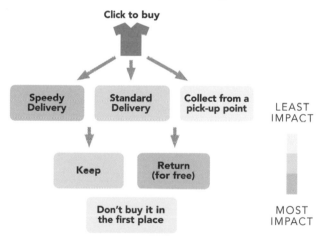

Click to buy

| Speedy Delivery | Standard Delivery | Collect from a pick-up point |

LEAST IMPACT

| Keep | Return (for free) |

Don't buy it in the first place

MOST IMPACT

Did you know?

International shipping constitutes approximately 3–4 percent of human-caused carbon emissions. By 2050, that figure is estimated to rise to 17 percent.

HOW DO THESE CHOICES AFFECT THE ENVIRONMENT?

Speedy delivery: As soon as you click on super-fast delivery, the business has made a promise to you that your package will arrive by the selected day. Speedy delivery will often mean that more, and emptier, vehicles are driving for longer hours of the day as they race against the clock to get to your address. Sometimes, if the item isn't in stock at the local warehouse, the company will have to fetch it from another warehouse further away—even via plane or ship, if necessary.

Standard delivery: This is much better than speedy delivery, but there are still negative implications, as the delivery vehicle is required to drive to your exact location, carrying potentially even the smallest, most insignificant purchase. Another problem with standard delivery is that you can't choose a day or time for your package to arrive. Since up to 40 percent of all deliveries are unsuccessful on the first try, further attempts waste even more miles. This is made worse still if, eventually, the customer has to drive somewhere to collect their package.

Collect from a pick-up point: As long as you are collecting your package on foot or the journey to the pick-up point is short, this is the most sustainable way to do online shopping.

Returns: About one-fifth of all online purchases are returned, mainly due to companies offering the service for free. Given this benefit, many consumers will order the product they want in a few sizes and colors, keep the one they like best, and send the rest back. This doubles the transportation carbon footprint of each purchase.

Keep it: If it's an item that's essential then great, but if you bought it because you were in need of some retail therapy, you've still increased your carbon footprint.

Don't buy it: Before every online purchase, consider it well. Will it definitely suit you? Will it 100 percent fit? Is it something you'll wear once and never again? Is it just another fad that'll sit at the back of a closet collecting dust? Will your life be different without it? Each time you go to purchase something, check in first with your eco-self.

Did you know?

More than 55 percent of consumers visit stores before buying online, helping to increase the average carbon footprint of shoppers dramatically. If you see something in a store that you really need, try to buy it then and there, rather than going online for a cheaper deal.

Approximately 30 percent of eBay's top 10,000 sellers and 40 percent of Amazon Marketplace sellers are located in China. Be mindful about which companies you buy from, as shipping a single product halfway across the world can multiply the size of its carbon footprint.

OUT AND ABOUT

After working hard through the week, it's natural to seek a little bit of relief in leisure activities and relaxing before doing it all again come Monday. One way to enjoy our free time is to go out exploring new places or returning to old haunts. But with all this to-ing and fro-ing—on top of the journeys you might take getting to work, transporting children, and visiting relatives—it's easy to forget what impact some of these activities are having on the planet.

THE CARBON FOOTPRINT OF CARS

It's difficult for some to imagine life without four wheels, but road vehicles are the biggest offenders of travel pollution, emitting 72 percent of the sector's carbon dioxide. Let's take a look at the damage they are causing:

In the US, passenger cars emit 59 percent of CO_2 emissions from all transportation.

Manufacturing a medium-sized car produces more than 18 tn (16,329 kg) of CO_2.

A medium-sized car emits, on average, 5 tn (4,535 kg)—the equivalent of four small cars—of CO_2 in fuel each year.

Platinum metals, which have the greatest environmental impact of all metals, are used to coat catalytic converters. In order to produce 1 lb (.45 kg) of these metals, thousands of pounds of CO_2 emissions are produced.

SUVs are the worst culprits for emissions, yet there has been a rapid increase of sales over the past ten years. From 2008 to 2018, the market has grown from 8 percent to 32 percent in Europe, and in America sales reached almost 70 percent of total car sales in 2018.

WAYS TO REDUCE YOUR CARBON FOOTPRINT

Start a work carpool: Ask around to find out who drives to work in your company and, if they live near you, organize ride shares. Check if your organization has implemented a plan, as you might be able to benefit from priority parking or other initiatives.

Work from home: If your job role allows this, request to work from home whenever possible to reduce commuting emissions.

Take public transit: If you commute to work by car, think about taking the train or a bus instead. It might even work out cheaper, as you are often able to buy weekly, monthly, or annual passes that are a fraction of the cost of single-day tickets.

Justify each car journey: Before you start the engine, ask yourself if you really need to make the trip. Could you walk or cycle, if you're just going a short distance? Could you combine your errand with other errands on a different day and reduce your outings to just one?

Keep your car empty: Extra baggage only increases the amount of fuel you use.

Get your car serviced annually: An engine that doesn't get enough TLC is at least 10 percent less economical.

Look after your car: Check the oil and tire pressures regularly, and keep them at the correct levels.

Drive slowly and smoothly: 50 mph (80 km/h) is said to be the most economical speed to drive at on a motorway, and limited braking can decrease fuel consumption by 40 percent.

Opt for a hybrid or electric car, if yours is on its last legs. As well as being environmentally friendly, it could also be a savvy choice to make, since many manufacturers are planning to phase out diesel and gasoline cars within the next decade.

THE CARBON FOOTPRINT OF PLANES

Globally, we have increased our air travel by 300 percent since 1990, and taking just one round-trip flight per year could amount to the same carbon footprint of a person who lives in a developing country for a *whole* year. Consider how much CO_2 and its equivalents are produced for various round-trip flights that start from London, UK:

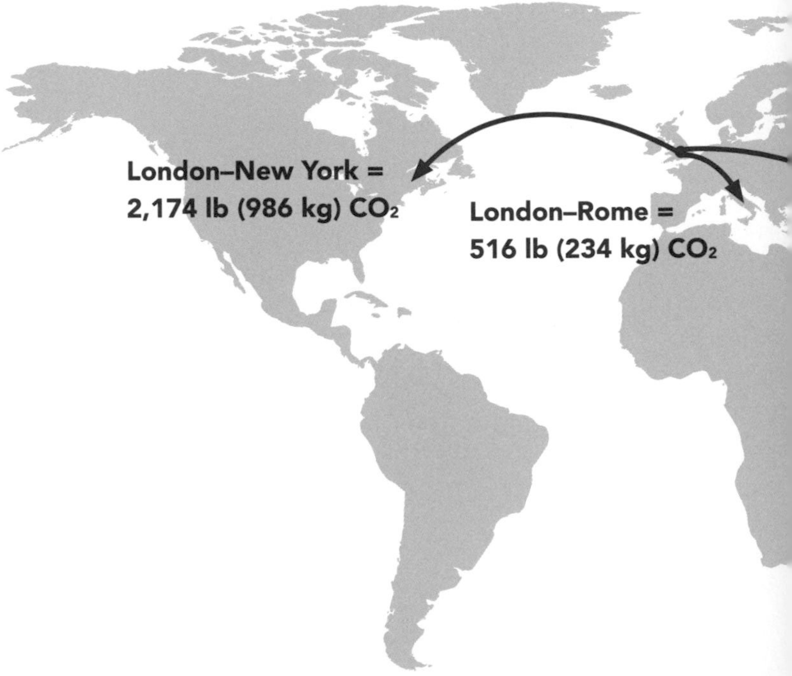

London–New York = 2,174 lb (986 kg) CO_2

London–Rome = 516 lb (234 kg) CO_2

**London–Perth =
6,951 lb (3,153 kg) CO_2**

WAYS TO REDUCE YOUR FLIGHT'S CARBON FOOTPRINT

If you have to fly for work purposes, always consider how necessary it is. Could you instead Skype or have a conference call with the people you are scheduled to meet? If it's an important meeting, see if you can arrange to see other clients within the area, so as not to double up on trips.

Research which airlines have smaller carbon footprints and fly with them.

Fly from your nearest airport and use public transportation to get there to minimize the emissions of driving.

Fly economy class: Even if your company is paying and can afford a more expensive ticket. Flying on a plane with luxurious, space-hogging first-class areas increases your footprint by three times.

Travel light: A heavier plane means more fuel is used.

Fly direct: Don't take multiple flights to a destination just because it's cheaper. Not only are you traveling farther, you are also multiplying your takeoffs and landings, which are the parts of the journey that use up the most fuel.

Have more staycations and fewer weekend trips: Getaways are great, until you think about them from an environmental point of view. Unlike choosing one main vacation that lasts for one or two weeks, it's likely that you'll book multiple trips a year in order to get your vacation fix.

WHAT IS THE BEST MODE OF TRANSPORTATION?

Of course, walking, running, cycling, scootering, and skating are your best ways of getting around with a guilt-free carbon footprint conscience, but if you rely on a faster form of transportation to make your daily journeys, what's the best way to travel?

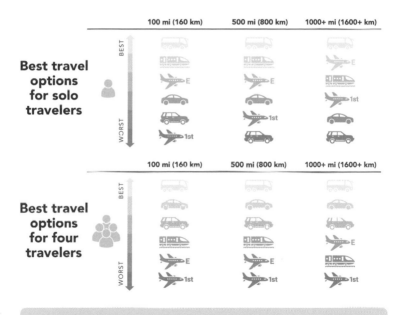

If you regularly travel across Europe, check out your train travel options as a greener way to travel than by plane. It can reduce your carbon footprint by up to 80 percent.

THE TRUTH ABOUT CRUISE SHIPS

It's often thought that cruise ships are a much greener way to travel than by plane. However, recent evidence shows that there's a darker side to these floating kingdoms.

In 2018, 76 out of 77 cruise ships that were inspected by German watchdog Nabu were found to use toxic heavy fuel oil (HFO), often identified as a "worse-case substance."

The air quality on the deck of cruise ships is comparable to the air quality of some of the most polluted cities in the world.

One cruise ship emits as much pollution as seven hundred trucks.

A passenger on a week-long cruise produces the same amount of carbon emissions as the average person would over eighteen days on land.

Sewage, trash, sulfur, and other dangerous pollutants from cruise ships are being irresponsibly disposed of into the oceans.

IN THE GARDEN

If you have a garden, it's easy to assume that your green space is a no-carbon-emissions zone. However, on closer inspection, the typical garden isn't as natural as it seems. From the fertilizer and weed killer to the gadgets and materials you use to make your garden a perfectly pruned haven, this section will take a look at some surprising facts, as well as ways to make your outdoor living greener.

THE CARBON FOOTPRINT OF GARDENS

On average, Americans use 88 million pounds (approx. 40 million kilograms) of fertilizer a year. Inorganic fertilizer produces nitrous oxide, a greenhouse gas about three hundred times more effective at trapping heat than carbon dioxide.

In the UK, 69 percent of peat compost sold is used for household gardening. Like forests, peatlands are rich stores of carbon dioxide, and losing just 5 percent of them in the UK would be the equivalent to gaining a year's worth of the country's carbon emissions.

The lawn care industry produces 13 billion pounds (approx. 6 billion kilograms) of toxic pollutants per year.

Studies have revealed that, on average, 2.5 acres (a hectare) of maintained lawn in Nashville, Tennessee, produces greenhouse gases equivalent to 1,537–5,386 lb (697–2,443 kg) of carbon dioxide a year.

Gas lawnmowers are harmful to the environment and our health, producing 1 lb (.45 kg) of carbon monoxide and several ounces/grams of methane, hydrocarbons, nitrogen oxides, and smoke particles for every hour they are used.

In 2011, Americans sent 15.9 million tons (over 31 billion pounds) of grass clippings to landfills, creating pollution via transportation and forcing the grass to decompose unnaturally, thus releasing high levels of methane.

WAYS TO REDUCE YOUR GARDEN CARBON FOOTPRINT

Plant trees, shrubs, and flowers that are native to your area—you will be dramatically reducing the number of miles/kilometers the items have to travel. Also, maintenance is lower, as they'll be in their usual habitat, and they will help to save carbon from entering the atmosphere.

Reduce your use of power tools: If your partner asks why the grass is getting so long, blame this book. Reducing the amount of times you use electric/gas-powered tools will reduce the amount of fossil fuels you burn and will make your garden thrive naturally.

Don't bag up grass cuttings—sending them to a landfill in a plastic bag means that they can't break down naturally. Instead, recycle your clippings and lay them directly onto your cut grass to act as natural fertilizer, compost them, or find out if you can take them to your local recycling center.

Stop using harmful fertilizers: Chemical fertilizers increase air and water pollution, as well as soil acidification. If you have to use fertilizer, buy organic instead.

Try growing your own food: From planting fruit trees to growing herbs and root vegetables, a small area in your garden dedicated to this has multiple environmental and economic benefits.

Create a resourceful garden: Pay back nature by making your garden a haven for animals and insects. Buy a bird bath and put out bird seed; put up a bee hotel; plant an abundance of flower species to attract a wide range of insects and animals. To make use of an excess of natural materials, install a rain barrel to catch rainwater, and start a compost heap for garden waste and vegetable matter from your kitchen.

HOW TO MAKE A COMPOST HEAP

Instead of sending your kitchen and garden waste to a landfill, make your own nutritiously rich compost heap:

Buy or make your own compost bin (the bigger, the better).

Find an area of your garden that is shaded; placing the compost bin straight on top of soil is best.

Add up to 50 percent of "green" material—including food waste (not meat!), grass clippings, leafy plants, weeds—and the rest "brown" material—e.g., prunings, hedge trimmings, leaves, paper or card (in small pieces), plant stems, straw, and mulch.

Turn the waste frequently (every week) to allow air into it.

You should get good results within between six months and two years.

Composting—top tips

- Keep the heap moist and loose.

- If the heap becomes slimy and smelly, there is too much "green" material (like grass cuttings) and it is too moist—counter the problem by adding more "brown" material (like dead leaves) and covering the heap to protect it from the rain.

- If the heap is too dry and doesn't look like it's breaking down, try adding more green waste to it and a little water.

- If your heap is attracting flies, make sure your kitchen waste is well hidden under other material.

TOP TEN WAYS TO CUT YOUR CARBON CONSUMPTION

Remember to:

1. Stop eating (or eat much less) meat.

2. Buy local and organic produce.

3. Walk, cycle, or take public transportation.

4. Make your holidays "staycations."

5. Reduce your use of energy-draining appliances.

6. Stop buying items that you don't need.

7. Avoid purchasing "fast fashion."

8. Recycle and reuse.

9. Make the holidays memorable, not material.

10. Strive to implement a zero-waste policy in all areas of your life.

CLIMATE
CHANGE
CRUSADERS

> **66** The more people invest in fossil fuel companies, the more these companies will exploit fossil fuels. **99**
>
> **Bob Watson**

- -

A large part of a business's success depends on the demand for its product. That is to say, we as consumers have a huge influence on what they manufacture and how they do it. Let's make a positive impact on the environment by boycotting the culprits of climate change and supporting the pioneers. We still have a very long way to go, but this chapter champions the positive changes that have been made so far.

INDIVIDUALS WHO ARE CHANGING THE STORY

In 2019, **633 divers** embarked on an underwater clean-up, helping to remove 1.7 tn (1,542 kg) of waste from the ocean floor in Florida and becoming Guinness World Record title holders.

Bachendri Pal, the first Indian woman to climb Mount Everest, has removed, with the help of a team of volunteers, 62 tn (over 56,000 kg) of waste from the Ganges river.

Jadav Payeng, AKA the Forest Man of India, has grown 1,359 acres (550 hectares) of forest by planting one tree a day for forty years. It is bigger than Central Park and home to hundreds of animals, including elephants, reptiles, and boars.

People in the **community of Kwinana**, Australia, put their heads together to come up with a way of capturing waste in their water cycle. They installed two drainage nets in a reserve and collected 816 lb (370 kg) of trash in just four months.

Swiss entrepreneur, businessman, and philanthropist **Hansjörg Wyss** has pledged $1 billion to help conserve 30 percent of the planet's surface by 2030.

In 2015, environmentalist **Afroz Shaz** began picking up waste on a local beach in Mumbai. Since then he has picked up 11.7 million pounds (approx. 5.3 million kg) of waste with the help of a 1,000-strong army he recruited over the years.

British investor **Jeremy Grantham** has long advocated for socially responsible investing while warning the finance community of the dangers posed by climate change. And in 2019, he pledged to give 98% of his personal wealth, over 1 billion dollars, to fight climate change.

YOUNGER PIONEERS

Xiuhtezcatl Martinez was just six when he realized the impact of human activity on the planet. Now he is a spokesperson at important climate change events and the youth director of conservation organization Earth Guardians.

At the age of nine, **Ridhima Pandey** filed a complaint against the Indian government to get them to take climate change more seriously. She has since written a fifty-two-page petition and become a spokesperson at numerous environmental events.

Timoci Naulusala was just twelve when he took to the stage at the UN's annual conference on climate change, telling world leaders about a cyclone that devastated his village.

Now a household name, **Greta Thunberg** got the media's attention when, in August 2018, at just fifteen years old, she began missing school to stand outside the Swedish parliament in an effort to get the government to act more vigorously in fighting against climate change. Since then, she has addressed the UN Climate Action Summit in New York and received honors and awards, including being named one of the most influential people of 2019 by *Time* magazine.

BECOME AN ACTIVIST

66 Never doubt that a small group of thoughtful, committed citizens can change the world. Indeed, it is the only thing that ever has. **99**

Margaret Mead

- -

Have the stories in the previous chapter inspired you to not only make your own lifestyle changes, but also try to encourage others—whether that's individuals, businesses, or even governments—to get on board the (100-percent renewable energy) eco-train? It's easier to do than you might think. Just take a look at some of the ways you can become involved in the fight against climate change.

HOW TO BECOME AN ENVIRONMENTAL ACTIVIST

- Keep up to date with the latest findings and news stories, and open up discussions about them with people you see.

- Attend or organize local events, such as beach clean-ups and environmental talks.

- Volunteer at local conservation charities.

- Contact local authorities about their climate change policies and demand change.

- Join a political lobbying network.

- Sign petitions you believe in.

- Sign up to green charities' e-newsletters.

- Donate to these charities.

- Educate children (your own, or give a presentation at a school) on climate change and make them aware of how to be good to the planet.

- Participate in demonstrations.

- Get your voice and opinion heard in any law-abiding way you can.

CONCLUSION

The future is never set in stone and, despite serious worries about global warming wreaking havoc on the planet, there is still time to repair some of the damage we have caused. While there are naysayers who deny the severity of the issue, as well as CEOs and global leaders who are indifferent to the emergency, you still have to believe that you can make a difference, however small it might be. Because, you see, if every individual did this, then each person's small difference would add up to a pretty big one. So, let's pay it back to our planet that has been our home and protector for all these years. The future is what we make it.

IMAGE CREDITS